THE POETRY OF HYDROGEN

The Poetry of Hydrogen

Walter the Educator™

SKB

Silent King Books a WhichHead Imprint

Copyright © 2023 by Walter the Educator™

All rights reserved. No part of this book may be reproduced in any manner whatsoever without written permission except in the case of brief quotations embodied in critical articles and reviews.

First Printing, 2023

Disclaimer
This book is a literary work; poems are not about specific persons, locations, situations, and/or circumstances. This book is for entertainment and informational purposes only. The author and publisher offer this information without warranties expressed or implied. No matter the grounds, neither the author nor the publisher will be accountable for any losses, injuries, or other damages caused by the reader's use of this book. The use of this book acknowledges an understanding and acceptance of this disclaimer.

dedicated to all the chemistry lovers,
like myself, across the world

CONTENTS

Dedication v

Why I Created This Book? 1

One - Cornerstone Of Science 2

Two - Universe's Core 4

Three - Unlock The Unknown 6

Four - Fertile Ground 8

Five - Proteins To Fuels 10

Six - Catalyst For Progress 12

Seven - Destruction And Mankind 14

Eight - Respect And Care 16

Nine - The Key 18

Ten - Element Divine 20

Eleven - Devoid Of Divisions 22

Twelve - Perfect Blend 23

Thirteen - Hydrogen's Potential		25
Fourteen - Versatile		26
Fifteen - Harness Its Power		27
Sixteen - Wherever You Roam		29
Seventeen - Mighty Force		31
Eighteen - Our Ally		33
Nineteen - Brighter Tomorrow		35
Twenty - Beacon Of Hope		36
Twenty-One - Limitless Scope		38
Twenty-Two - Environment's Betterment		40
Twenty-Three - Leads The Way		42
Twenty-Four - Knows No Bounds		44
Twenty-Five - Harmony And Growth		46
Twenty-Six - Change We Need		48
Twenty-Seven - Hydrogen, Oh Hydrogen		50
Twenty-Eight - Humble Element		52
Twenty-Nine - Light And Pure		54
Thirty - Combating Pollution		56
Thirty-One - Greener Sight		58
Thirty-Two - Gift From The Heavens		60

Thirty-Three - We Proclaim 62

Thirty-Four - Each Molecule 64

Thirty-Five - Catch Fire 66

Thirty-Six - Sets Us Free 68

About The Author 70

WHY I CREATED THIS BOOK?

This Hydrogen-themed poetry book offers a blend of science, art, education, symbolism, and environmental consciousness, making it a compelling and thought-provoking work. It can inspire readers to appreciate the beauty of science, ponder the mysteries of the universe, and contemplate the importance of sustainable energy solutions.

ONE

CORNERSTONE OF SCIENCE

In the realm of atoms, a star is born,
A humble element, Hydrogen, is sworn.
With but a single proton, it dwells,
A tale of simplicity, this element tells.

In the core of the sun, where fusion ignites,
Hydrogen dances, its brilliance takes flight.
Bound by forces, strong and weak,
It powers the cosmos, both mild and meek.

Hydrogen, the fuel of celestial fire,
It dances with helium, a cosmic desire.
From swirling nebulae to galaxies afar,
It shapes the cosmos, like a guiding star.

In the depths of oceans, where life does dwell,
Hydrogen bonds, a tale to tell.

With oxygen it weaves, a liquid embrace,
Creating the elixir, sustaining life's grace.

But beyond its beauty, in the explosive realm,
Hydrogen reveals its destructive helm.
Unleashed in fury, with power untamed,
It brings devastation, leaving none unclaimed.

Yet, in the hands of scientists, its power is tamed,
Harnessing its energy, as humans have aimed.
From fuel cells to rockets, its potential so vast,
Hydrogen propels us to a future so vast.

Oh, Hydrogen, element of dreams,
A building block, it seems,
From the vast expanse of space,
To the smallest atom's embrace.

Unique and versatile, this element divine,
Hydrogen, in the realm of chemistry, does shine.
Its story unfolds, a tale ever new,
A cornerstone of science, forever true.

TWO

UNIVERSE'S CORE

In the heart of the sun, a fiery dance,
Hydrogen, the element, takes its chance.
A single proton, a humble start,
Yet from its depths, the cosmos does impart.

From nebulae to galaxies afar,
Hydrogen's presence, a celestial star.
Fuel for the stars, their brilliant light,
A cosmic symphony, a wondrous sight.

On Earth, it weaves its magic spell,
In every molecule, it loves to dwell.
Water's companion, life's elixir,
Hydrogen's touch, a constant mixer.

Within the atoms, it forms a bond,
A bridge between elements, strong and fond.
In molecules, it brings life's zest,
Hydrogen, the element, we are blessed.

But beware, dear souls, the hidden might,
For Hydrogen can unleash a fearsome fight.
In its fusion core, atomic power,
A force unleashed, in a single hour.

Yet in its duality, lies a key,
For Hydrogen, the element, holds the key.
To harness energy, a dream so bright,
A future propelled by Hydrogen's might.

In fuel cells, it powers the land,
A cleaner tomorrow, within our hand.
From cars to homes, a sustainable choice,
Hydrogen whispers, a hopeful voice.

So let us embrace this versatile friend,
In chemistry's realm, it knows no end.
Hydrogen, the element, let us adore,
For it holds the secrets of the universe's core.

THREE

UNLOCK THE UNKNOWN

In the vast expanse of cosmic heights,
Where stars ignite their eternal lights,
There lies a secret, a shimmering gem,
A humble element, Hydrogen, a true chem.

From the depths of space, it does emerge,
A timeless dance, a celestial surge,
For in the stars, it finds its birth,
Nuclear fusion, a cosmic mirth.

With boundless might, it fuels the Sun,
A fiery furnace, where life begun,
In its core, where atoms collide,
Hydrogen's power, cannot hide.

But not just in the heavens above,
On Earth, it manifests its love,

In every drop of water's grace,
Hydrogen weaves its sacred trace.
    Within the H2O, it finds its place,
A molecule of life, full of grace,
Quenching thirst, sustaining all,
Hydrogen's presence, never small.
    Yet, beyond the realms of water's flow,
Hydrogen's secrets begin to show,
In chemistry's realm, it takes its stand,
A building block, so grand.
    From fuels that burn, to acids that bite,
Hydrogen's versatility shines bright,
A reagent, a catalyst, a chemical key,
Unlocking reactions, with chemistry.
    But beware, for in its volatile form,
Hydrogen can unleash a storm,
Explosions fierce, with fiery might,
A cautionary tale, in the darkest night.
    And yet, from destruction, hope can arise,
Harnessing Hydrogen, a future that lies,
In clean energy, a beacon of light,
A solution to our planet's plight.
    So, let us cherish this element grand,
From the cosmos to the palm of our hand,
Hydrogen, a key to unlock the unknown,
In the vast universe, we shall be shown.

FOUR

FERTILE GROUND

In the depths of the cosmic expanse,
Where stars dance and galaxies prance,
There lies a wondrous element, pure and light,
Hydrogen, the foundation of celestial might.

From the birth of time, it has been there,
A silent witness to the universe's affair,
In stellar nurseries, it ignites the spark,
Binding atoms together, creating a celestial ark.

In the heart of the sun, it fiercely burns,
Releasing energy, the cosmos it churns,
A fusion of atoms, a radiant blaze,
Hydrogen's power, the sun's eternal praise.

But on Earth's surface, it takes a different form,
Bound to oxygen, it embraces a new norm,
Water, the elixir of life, it becomes,
Hydrogen's embrace, where existence thrums.

In chemistry's realm, it shows great might,
A versatile element, shining so bright,
Fueling reactions, it dances with grace,
Creating bonds, a molecular embrace.

Yet, its power lies not just in chemistry's hold,
For it can be harnessed, a story yet untold,
A promise of clean energy, a renewable source,
Hydrogen's potential, a sustainable force.

From fuel cells to vehicles that roam,
Hydrogen whispers of a greener home,
A world free from pollution's cruel toll,
A future where clean energy takes control.

So let us marvel at Hydrogen's grace,
An element woven into time and space,
From the cosmos to Earth's fertile ground,
Hydrogen, a powerful force that astounds.

FIVE

PROTEINS TO FUELS

In the depths of the ocean, where life begins,
Lies a secret hidden, where wonder begins.
Hydrogen, the essence of water's embrace,
A molecule of life, in every drop's trace.

From the shimmering rivers to the rain-soaked ground,
Hydrogen dances, its magic profound.
It bonds with oxygen, creating the flow,
A symphony of molecules, a ballet of H2O.

But Hydrogen's powers, they go far beyond,
In the realm of chemistry, it's a virtuoso bond.
A fuel for the stars, a cosmic delight,
In the sun's core, it ignites with pure light.

With electrons in tow, it dances with grace,
Hydrogen, the building block, in every space.

From proteins to fuels, it weaves its design,
A versatile element, forever benign.
 And as we dream of a world clean and bright,
Hydrogen whispers of renewable might.
A promise of energy, sustainable and true,
A future where Hydrogen will carry us through.
 So let us embrace this element divine,
For Hydrogen's potential, it can surely align.
From the depths of the ocean to the sun's golden hue,
Hydrogen, the key to a future anew.

SIX

CATALYST FOR PROGRESS

In the realm of chemistry's dance,
Hydrogen, a versatile trance.
Its tiny atom, so light and pure,
Holds power and promise to endure.

A single proton, a lone electron,
Hydrogen's essence, a gem to reckon.
With carbon, oxygen, or nitrogen's might,
It forms bonds, weaving compounds so bright.

In flames it crackles, with oxygen it ignites,
A fiery partnership, burning so bright.
In water, it floats, invisible and clear,
Hydrogen whispers, its secret held dear.

But beyond the lab, its potential lies,
A promise of greener skies.

As fuel for the future, it beckons us near,
A clean energy source, without a fear.

Hydrogen, the element of wonder,
Unleashing power, pulling us under.
From stars it's born, in molecules it's found,
A building block, in the cosmos profound.

So let us embrace this element divine,
Harness its energy, let it shine.
Hydrogen, a gift, both small and grand,
A catalyst for progress, in our hands.

SEVEN

DESTRUCTION AND MANKIND

In a world of atoms, small and bright,
Hydrogen reigns, a paradox of might.
With a single proton, a gleaming shell,
It holds the secrets, we long to tell.

Hydrogen, a spark of cosmic birth,
Flares to life, a force upon this Earth.
But beware, for in its fiery core,
Lies the power to destroy and more.

In bombs and stars, its fury unleashed,
Hydrogen wreaks havoc, power increased.
Explosions thunder, skies ablaze,
Leaving destruction in its fiery haze.

Yet, amidst the chaos, a glimmer of hope,
Hydrogen whispers, a promise to cope.

For in its essence, a power untamed,
A source of energy, yet to be named.
    From fuel cells to fusion dreams,
Hydrogen's potential, it truly seems,
To light our homes, power our cars,
A future where clean energy soars.
    So let us harness this element divine,
Unleash its power, a sustainable sign.
For in Hydrogen's dance, we find,
A balance between destruction and mankind.

EIGHT

RESPECT AND CARE

In the realm of atoms, a king stands tall,
Hydrogen, the element, the first of them all.
A solitary proton, a single electron,
This humble element, its power yet unspoken.

With a heart of stars, it lights up the night,
A cosmic dance, a celestial delight.
In fusion's embrace, it ignites the sun's fire,
A radiant force, an eternal desire.

Hydrogen, the fuel for reactions untamed,
Combining with others, its power unchained.
It joins with oxygen, a watery bond,
Quenching the thirst of a parched land beyond.

From fiery explosions to gentle caress,
Hydrogen's nature, a delicate finesse.
A volatile presence, it whispers and roars,
Unleashing its might, breaking down all doors.

But its potential goes beyond its might,
A promise of progress, a future so bright.
A clean energy source, a beacon of hope,
A sustainable world, where green dreams can elope.
　So let us embrace this element sublime,
Harness its power, for it is not a crime.
With responsibility, we'll pave the way,
To a greener future, where Hydrogen holds sway.
　Hydrogen, the element, so versatile and grand,
A key to the future, in its mighty hand.
Let us tread lightly, with respect and care,
For in Hydrogen's embrace, lies a world we can share.

NINE

THE KEY

In the realm of atoms, a star is born,
A tiny particle, a spark that adorns.
Hydrogen, the element of boundless might,
A dance of protons, electrons so light.

A single proton, a nucleus so small,
Yet holds the power to ignite us all.
A silent force, a universe untamed,
In its simplicity, a power untamed.

From the stars above to the Earth below,
Hydrogen's secrets, it begins to show.
In the chemistry lab, reactions unfold,
A symphony of elements, stories untold.

With oxygen it bonds, a fiery embrace,
Creating water, life's eternal grace.
But beyond its role in nature's design,
Hydrogen dreams of a future divine.

A clean energy source, it yearns to be,
A catalyst for progress, for all to see.
From fuel cells to rockets, it takes us high,
To a greener future, it helps us fly.

Yet with power comes responsibility,
To harness its might with sensitivity.
For Hydrogen's flames can consume and destroy,
But with careful tending, it brings us joy.

In the hands of mankind, the choice is ours,
To harness Hydrogen's infinite powers.
To walk the tightrope of balance and grace,
And embrace its potential, a brighter space.

So let us honor this element so grand,
Hydrogen, the key to a future unplanned.
In its versatile nature, a promise we find,
To create a world that's gentle and kind.

TEN

ELEMENT DIVINE

In the realm of chemistry, a star is born,
A versatile element, Hydrogen adorned.
With a single proton, and an electron too,
It holds the promise of a future anew.

Hydrogen, the lightest of them all,
Invisible and humble, yet standing tall.
It fuels the stars that light up the night,
A cosmic dance of pure delight.

But on Earth, its potential is vast,
A clean energy source that can outlast.
From fuel cells to power our cars,
To heating homes and reaching for the stars.

Hydrogen, a gift from the universe above,
A catalyst for progress, a symbol of love.
Its atoms intertwine, forming bonds so strong,
A testament to its potential all along.

So let us embrace this element divine,
Harness its power, let it brightly shine.
With Hydrogen as our guiding light,
We'll pave the way to a future bright.

ELEVEN

DEVOID OF DIVISIONS

A power untamed, it does disguise.
A tiny atom, so light and small,
Yet holds the key to power for all.
With boundless potential, it yearns to be,
The fuel of progress, clean and free.
In stars it dances, igniting the sky,
A cosmic force that cannot deny.
From sunlit fields to bustling streets,
Hydrogen's promise, the world it greets.
As cars roll on, with zero emissions,
A future bright, devoid of divisions.
Oh, Hydrogen, a beacon of hope,
A chance to heal, a wider scope.
With every molecule that takes flight,
We steer our world towards a brighter light.

TWELVE

PERFECT BLEND

In the realm of energy, a star so bright,
Hydrogen, the element, shines with might.
A power untamed, a force so pure,
Unleashing its potential, to endure.

From the depths of stars, it does arise,
A fuel for the future, in our skies.
Clean and abundant, it holds the key,
To a world that's greener, for you and me.

With particles dancing, in bonds so strong,
Hydrogen sings its clean energy song.
A catalyst for progress, it lights the way,
A promise of a brighter, sustainable day.

From fuel cells to rockets, it takes its flight,
Hydrogen's power, an awe-inspiring sight.
In engines and turbines, it fuels our dreams,
A cleaner, brighter future, it gleams.

So let us embrace this element divine,
Harness its power, let it intertwine,
With nature's harmony, in perfect blend,
Hydrogen, our ally, on which we depend.

THIRTEEN

HYDROGEN'S POTENTIAL

Invisible and light, Hydrogen's might
Is in its power to fuel the fight
Against pollution and climate change
Clean energy source, a global range
From cars to homes, to industry's might
Hydrogen's potential, a shining light
A renewable fuel, a sustainable choice
A cleaner future, we can all rejoice
So let's embrace this element so pure
And fuel progress, of that we can be sure
Hydrogen, the building block of life
A greener world, without any strife.

FOURTEEN

VERSATILE

Hydrogen, the element of one,
The simplest of them all,
A colorless gas that's often spun,
To power up the world.

From stars above to Earth below,
Its presence is everywhere,
A building block of life we know,
And fuel for progress here.

But now we see its greater worth,
As we seek a greener way,
To power up our homes and earth,
And keep pollution at bay.

Hydrogen, so versatile,
A clean energy source so bright,
With potential to reconcile,
Our planet's future in sight.

FIFTEEN

HARNESS ITS POWER

In a world where progress blooms,
A whisper of promise, Hydrogen looms.
A humble element, light and pure,
With potential to reshape our future.

From the core of stars, it takes its birth,
A cosmic dance, a celestial mirth.
In every atom, it silently resides,
Fueling the universe, where wonder resides.

But here on Earth, its purpose unfolds,
A clean energy source, a story untold.
With molecules of hope, it fills the air,
A catalyst for change, beyond compare.

Hydrogen, the champion in our plight,
Combating pollution, shining so bright.
Its gentle touch, a balm for the Earth,
A savior in the fight against climate's dearth.

In the heart of engines, it ignites the flame,
Powering our dreams, a sustainable game.
No fumes, no smoke, just water as it goes,
A greener world, its legacy it bestows.

From transportation to homes and more,
Hydrogen's versatility we adore.
Electrifying our paths, with zero emission,
A fuel of the future, a green transition.

So let us embrace this element divine,
Harness its power, let our world align.
For Hydrogen, a solution we find,
A cleaner, brighter future, intertwined.

SIXTEEN

WHEREVER YOU ROAM

In the realm of atoms, a star is born,
A tiny particle, Hydrogen is sworn.
With just one proton, it stands alone,
The lightest element, a beauty of its own.

From the depths of space to the Earth's delight,
Hydrogen's potential shines so bright.
A clean energy source, it holds the key,
To a world free from pollution, for you and me.

In the heart of the sun, it fuels the fire,
Bringing warmth and light, never to expire.
In fuel cells, it powers our dreams,
Unleashing a future where everything gleams.

Hydrogen, oh Hydrogen, so versatile and pure,
You hold the power to create a greener world for sure.

From powering cars to heating our homes,
You make our lives better, wherever you roam.
 So let us embrace this element divine,
Harness its power, let our dreams intertwine.
With Hydrogen as our ally, we'll pave the way,
To a future where sustainability holds sway.

SEVENTEEN

MIGHTY FORCE

In the realm of atoms, a mighty force,
Lies an element, Hydrogen, a powerful source.
Burning bright, in celestial skies,
It holds the secrets of our future's rise.

Hydrogen, the lightest of them all,
A colorless gas, ready to enthrall.
With just one proton, and an electron's dance,
It carries the potential for a clean energy advance.

From the depths of stars, it was born,
A testament to the universe's endless form.
Now on Earth, it holds a different role,
A solution to pollution and the climate's toll.

Hydrogen, the fuel of a greener age,
Powering dreams on a boundless stage.
With its combustion, no toxic smoke,
Only water vapor, as it gracefully spoke.

In vehicles, it holds the key,
To a world that's clean and pollution-free.
Cars that run on hydrogen's might,
Gently whispering through the night.

A renewable source, with endless supply,
Harnessing its power, we can't deny.
From wind and solar, it can be made,
A beacon of hope, we need to crusade.

Hydrogen, the element of change,
In our fight against a planet's range.
With its potential, so vast and grand,
Let's embrace it, hand in hand.

So let us dream, of a future so bright,
Where Hydrogen's power takes flight.
A world that's sustainable, clean, and pure,
Where Hydrogen's potential will endure.

EIGHTEEN

OUR ALLY

In the realm of atoms, a star was born,
A humble element, Hydrogen adorned.
With a single proton, it shines so bright,
A beacon of hope, a source of light.

From the vast cosmos, to our earthly abode,
Hydrogen whispers secrets, yet untold.
Its potential, a gift, we can't ignore,
To shape a future, worth fighting for.

Oh Hydrogen, you hold the key,
To a world that's clean and pollution-free.
With your power, we can turn the tide,
And in harmony, with nature, we can reside.

No more smog-filled skies, no more toxic air,
Hydrogen, you're the answer, we declare.
A renewable fuel, with no carbon trace,
A symbol of progress, a saving grace.

In fuel cells, you dance, with electrons in flight,
Powering our dreams, with pure delight.
From cars to homes, and everything between,
Hydrogen, you're the future we have seen.

So let us embrace you, and harness your might,
Unleash your potential, both day and night.
Together we'll build a greener domain,
Hydrogen, our ally, forever to remain.

NINETEEN

BRIGHTER TOMORROW

At the heart of the sun, Hydrogen burns bright
A fusion of atoms, a celestial light
On Earth, it holds promise, a clean energy source
To combat pollution and climate change, of course
But Hydrogen is more than just fuel for the fire
Its versatility and power, it can inspire
From powering homes to fueling cars
Hydrogen takes us far
Let us embrace its potential, its endless supply
A greener future, we can't deny
Hydrogen, the element of change
Fueling progress and sustainability, in range
So let us harness its power, its energy so pure
And create a world that is green and secure
Hydrogen, the key to a brighter tomorrow
A world in harmony, with nature to borrow.

TWENTY

BEACON OF HOPE

In a realm of atoms, pure and bright,
A star-born element, gleaming with light.
Hydrogen, the cosmic fuel of the skies,
Unleashing power as it soars and flies.

From the depths of stars, it takes its birth,
A fusion dance, a celestial mirth.
Burning bright in the cosmic dance,
Hydrogen's energy, a dazzling trance.

But beyond the heavens, its potential lies,
To fuel a world, where progress never dies.
A clean energy source, it holds the key,
To combat pollution, set the planet free.

Hydrogen, the emissary of a greener dawn,
A remedy for a world that's been torn.
Harnessing its power, we can strive,
To create a future where all can thrive.

From cars to planes, and everything between,
Hydrogen's versatility, a sight unseen.
Powering nations with sustainable might,
A beacon of hope, shining so bright.

So let us embrace this element divine,
Unlocking a future where we all can shine.
Hydrogen, the catalyst for a world renewed,
A cleaner, brighter world that we can pursue.

TWENTY-ONE

LIMITLESS SCOPE

In the realm of atoms, a star's secret brew,
A molecule of promise, a world anew,
Hydrogen, the essence of cosmic might,
Unleashing its power, shining so bright.

From the sun's core, its energy flows,
Igniting the heavens, a celestial prose,
A fusion of particles, a radiant dance,
Hydrogen's embrace, a cosmic romance.

But beyond the skies, on Earth's fertile ground,
Hydrogen's potential, still to be found,
A clean energy source, it holds the key,
To combat pollution, set the planet free.

In fuel cells it thrives, a silent force,
Powering our lives, with minimal discourse,
No emissions, no smoke, no toxic fume,
Hydrogen's touch, a cleaner world to bloom.

From cars to homes, it lights up our days,
A renewable flame, in countless ways,
Transportation transformed, no longer chained,
Hydrogen's embrace, a future unstrained.
　　So let us embrace this element divine,
Harness its power, let it brightly shine,
Hydrogen, the promise, the fuel of hope,
A greener future, with limitless scope.

TWENTY-TWO

ENVIRONMENT'S BETTERMENT

Hydrogen, a gas so light,
Makes up the stars shining bright,
But on Earth, it's often used,
To power machines, and we're amused.

Its potential is quite grand,
As a clean fuel, it takes a stand,
Against pollution, it can fight,
And bring a future that's bright.

Hydrogen, a versatile friend,
That can help our planet mend,
From cars to planes, it can power,
A sustainable future, it can empower.

So let's embrace this element,
And use it to our environment's betterment,

For Hydrogen holds the key,
To a greener world for you and me.

TWENTY-THREE

LEADS THE WAY

In the realm of molecules, a humble friend I find,
Hydrogen, the element, with a power so kind.
With its single proton, it dances through the air,
A beacon of potential, a promise so rare.

Oh Hydrogen, the fuel of a greener age,
A savior from pollution, our planet's fierce rage.
In the heart of the sun, you burn bright and strong,
A fusion of energy, where new worlds belong.

From the depths of the ocean, you rise to the sky,
A source of clean power, never asking us why.
With water as your byproduct, no harm you leave behind,
A testament to nature's wisdom, so divinely designed.

In cars and in buses, you power our way,
No fumes or emissions, just a clear, bright day.

From homes to factories, you light up our lives,
A renewable source, where hope forever thrives.
    Hydrogen, the element, with potential untold,
A key to our future, a story yet to unfold.
In the quest for sustainability, you lead the way,
A symbol of progress, shining brighter with each passing day.

TWENTY-FOUR

KNOWS NO BOUNDS

In a realm of atoms, so light and serene,
There lies Hydrogen, an element unseen.
With but one proton, it stands alone,
A jewel of the periodic table, brightly shone.

Beyond its simplicity, a power resides,
A force that can change the world's tides.
For Hydrogen holds a promise so grand,
To fuel our dreams with a greener hand.

In its nucleus, the fusion of stars,
A cosmic dance that breaks all bars.
Unleashing energies, pure and bright,
Hydrogen's potential, a captivating sight.

From distant galaxies to our humble Earth,
Hydrogen's power, a renewable birth.
In fuel cells and engines, it takes its place,
Driving us forward, with elegant grace.

No more pollution, no more strife,
Hydrogen brings a sustainable life.
A future where the air is clean,
And our planet's beauty, forever seen.
  So let us embrace this element divine,
And harness its power, a gift so fine.
For Hydrogen's strength knows no bounds,
A cleaner, greener world it propounds.

TWENTY-FIVE

HARMONY AND GROWTH

In the depths of the universe, so vast and wide,
There exists a tiny atom, filled with pride.
Hydrogen, the element of the stars,
With its simple structure, it travels far.

A single proton, a lone electron's dance,
Hydrogen holds the power of circumstance.
With bonds so strong and energy untamed,
It fuels the dreams that cannot be named.

Hydrogen, the key to a cleaner tomorrow,
Unleashing its potential, free from sorrow.
A source of power, pure and divine,
In its embrace, pollution will decline.

From the depths of the Earth to the skies above,
Hydrogen whispers of a world we can love.

It powers our cars, our homes, our dreams,
With each molecule, progress redeems.

    Hydrogen, the spark of a greener age,
A catalyst for progress, turning the page.
It lights the path to a sustainable fate,
A future where harmony and growth relate.

    So let us harness this element's might,
Embrace its potential, shining so bright.
With Hydrogen's promise, let us aspire,
To a world that's cleaner, where dreams never tire.

TWENTY-SIX

CHANGE WE NEED

In the realm of science, a star is born,
A humble element, yet a force to adorn.
Hydrogen, the spark of the universe's might,
Unveiling its secrets, unveiling its light.

From the depths of the cosmos, it emerges,
A promise of clean energy, as it surges,
A beacon of hope, a solution we seek,
To combat pollution, and make our world sleek.

Hydrogen, oh Hydrogen, you hold the key,
To a future that's green, for all to see,
With your potential unleashed, we can dream,
Of a world where pollution is no longer supreme.

As fuel for our cars, you'll take us far,
With zero emissions, a shining star,
Transportation transformed, a new dawn,
Where clean energy reigns, and pollution is gone.

In our homes, you'll power the way,
A source of warmth and light, day by day,
No more reliance on fossil fuels' strife,
With Hydrogen, we'll embrace a sustainable life.
　　So let us embrace this element so pure,
Harness its power, of that we are sure,
Hydrogen, the catalyst for change we need,
To pave the way for a cleaner world, indeed.

TWENTY-SEVEN

HYDROGEN, OH HYDROGEN

Hydrogen, oh Hydrogen,
A world of power you contain within
Fueling machines and engines alike
A future of clean energy, you strike
From the sun's rays you can be derived
With water as your source, you thrive
No pollutants, no emissions to regret
A sustainable world, we can beget
Oh Hydrogen, your potential so great
A greener world, you can create
Combatting pollution with every use
A cleaner future, we can produce
Let us embrace your power and might
A world of progress, shining bright
With Hydrogen as our guiding star

We can propel ourselves forward, far.

TWENTY-EIGHT

HUMBLE ELEMENT

In a realm of atoms, shining bright,
A humble element, pure and light.
Hydrogen, the essence of the stars,
Unleashing power from afar.

A key to a future, clean and bright,
Harnessing energy, day and night.
With every proton, every bond,
A revolution, a world beyond.

From the depths of oceans, it can rise,
To power homes and cities, without compromise.
A fuel of hope, a breath of air,
Hydrogen, a promise we all share.

Through silent whispers, it will roam,
Reducing pollution, creating a greener home.
A catalyst for change, it takes the lead,
Hydrogen, the answer we all need.

So let us embrace its potential, its might,
A beacon of hope, shining so bright.
For in Hydrogen, we find our way,
To a future where sustainability holds sway.

TWENTY-NINE

LIGHT AND PURE

In a world where progress is sought,
There lies a secret, yet to be taught.
A humble element, light and pure,
A potential solution, long endured.

    Hydrogen, the atom of endless might,
A catalyst for a future, shining bright.
In its core, a power untamed,
A promise of a world, unashamed.

    From distant stars, it does emerge,
A gift from the heavens, a cosmic surge.
Its simplicity hides a grand design,
A universe within, a treasure to find.

    With electrons dancing, a delicate ballet,
Hydrogen whispers, "I'll show you the way."

A fuel for dreams, a boundless source,
Unlocking possibilities, a new discourse.
   From sunlit skies to ocean's deep,
Hydrogen's potential, we shall reap.
Cleaner and greener, a sustainable quest,
A future where harmony can finally rest.
   So let us embrace this humble gas,
And build a world that will forever last.
For Hydrogen's gift, we must cherish and hold,
A brighter future, its story unfolds.

THIRTY

COMBATING POLLUTION

Hydrogen, oh Hydrogen,
Element of the lightest kind,
Invisible and abundant,
A future we must find.

From stars and suns it comes,
A power source yet untapped,
Clean and renewable,
A future we must adapt.

In fuel cells it can shine,
Powering machines and engines,
A cleaner, greener world,
A future that beckons.

Oh Hydrogen, our hope,
In combating pollution and more,

A sustainable future we seek,
With you, we can soar.

THIRTY-ONE

GREENER SIGHT

In depths of space, where stars reside,
A humble atom waits inside.
Hydrogen, the first and lightest,
A universe's primordial guest.

With boundless energy, it does ignite,
A fusion dance, a cosmic light.
Nurtured in the heart of mighty stars,
It gives birth to life from afar.

From sunlit skies to ocean deep,
Hydrogen's secrets we shall keep.
A fuel of dreams, a promise bright,
To power our world with radiant light.

In cleanest flames, it does ignite,
No smoke, no soot, just pure delight.
A renewable source, a hope untold,
Hydrogen's future, a tale of gold.

So let us harness its potent might,
Embrace the dawn, a greener sight.
Hydrogen, a friend in every sense,
A force for change, a future immense.

THIRTY-TWO

GIFT FROM THE HEAVENS

In the realm of science, a star is born,
A humble element, Hydrogen adorned.
With a single proton, it stands alone,
In the vast expanse, a mystery to be known.

Through fusion's fiery dance, it ignites,
A beacon of hope, a celestial light.
A fuel of the future, clean and pure,
Hydrogen, the element we must secure.

Unleashing its power, a renewable force,
A world transformed, a cleaner course.
With zero emissions, it takes flight,
In a world burdened by pollution's blight.

From the sun's embrace, we seek to borrow,
A gift from the heavens, a brighter tomorrow.

Hydrogen, the key to a greener domain,
A catalyst for change, a sustainable refrain.
　With every molecule, a promise untold,
Invisible and abundant, its secrets unfold.
Unlocking the potential, it holds within,
A solution to the problems we've been in.
　Hydrogen, oh Hydrogen, a beacon so bright,
Shining a path towards a future so right.
Let us harness your power, embrace your might,
And pave the way for a world bathed in light.

THIRTY-THREE

WE PROCLAIM

In the realm of atoms, a star is born,
Hydrogen, the mighty, stands alone.
A single proton, a world untold,
Unleashing power, for it to behold.

From the sun's embrace, it travels far,
A fuel for dreams, a shining star.
Through fusion's dance, it gives us light,
In cosmic depths, a brilliant sight.

But beyond the sky, it finds its way,
To Earth's domain, where it holds sway.
A promise of a future, clean and bright,
A source of energy, like endless light.

Hydrogen, the element of change,
Unleashing power, we can't estrange.
With zero emissions, it takes its stance,
A catalyst for a sustainable advance.

From cars to homes, it fuels our needs,
A silent hero, in nature's deeds.
A world transformed, where pollution's tame,
Hydrogen, the element, we proclaim.

THIRTY-FOUR

EACH MOLECULE

In the realm of atoms, a spark ignites,
A whisper of power that shines so bright.
Hydrogen, the element of endless might,
A key to unlock a future so right.

From the depths of stars, it first emerged,
A cosmic dancer, the universe surged.
With one proton, it gracefully swirled,
The simplest of elements, a precious world.

Hydrogen, the fuel of dreams untold,
In its embrace, a new story unfolds.
Its atoms dance, a symphony of hope,
A promise of clean energy, a scope.

Through fusion's embrace, it holds the key,
To a world of wonder, for you and me.
With boundless potential, it takes flight,
A beacon of progress, shining so bright.

No more reliance on fossilized past,
Hydrogen's promise, a future so vast.
A cleaner world, where pollution is no more,
Renewable energy, for all to explore.

Oh, Hydrogen, the element divine,
A catalyst for change, a gift so fine.
With each molecule, a step we take,
Towards a greener world, for nature's sake.

So let us embrace this element's grace,
And strive for a future, a sustainable place.
With Hydrogen's power, we shall rise,
To a world where harmony never dies.

THIRTY-FIVE

CATCH FIRE

In the heart of the smallest star,
A secret lies, both near and far.
Hydrogen, the element so pure,
Unveils its potential, bright and sure.

Within its atoms, a power untold,
A source of energy, brave and bold.
From the sun's fiery embrace it's born,
A catalyst for change, a new dawn.

Hydrogen, the fuel of dreams,
Unleashing power, in silent streams.
A clean embrace, a renewable might,
Shining a path, like stars in the night.

From cars to homes, it lights the way,
A beacon of hope in skies so gray.
With every spark, a pollution's end,
Hydrogen, our ally, our dearest friend.

So let us harness this gift we've found,
And spread its glory all around.
For in its essence, we shall aspire,
To build a future, where dreams catch fire.

THIRTY-SIX

SETS US FREE

In the realm of atoms, a star is born,
A humble element, its power untorn.
Hydrogen, the lightest of them all,
With boundless potential, it does enthrall.

A catalyst of change, it takes the lead,
Unleashing energy, fulfilling our need.
In fusion's embrace, it glows and ignites,
A source of power, like celestial lights.

From sunlit skies to the depths below,
Hydrogen's versatility, it does bestow.
In fuel cells and rockets, it takes flight,
A cleaner energy, shining ever so bright.

It bonds with oxygen, creating pure water,
A symbol of hope, a sustainable slaughter.
Through chemistry's dance, it sets us free,
A cleaner world, for all to see.

Hydrogen, the element of promise and might,
A hero in disguise, shining with light.
Unlocking a future, where pollution is no more,
With hydrogen's embrace, we'll open that door.

ABOUT THE AUTHOR

Walter the Educator is one of the pseudonyms for Walter Anderson. Formally educated in Chemistry, Business, and Education, he is an educator, an author, a diverse entrepreneur, and the son of a disabled war veteran. "Walter the Educator" shares his time between educating and creating. He holds interests and owns several creative projects that entertain, enlighten, enhance, and educate, hoping to inspire and motivate you.

> Follow, find new works, and stay up to date
> with Walter the Educator™
> at www.WaltertheEducator.com

www.ingramcontent.com/pod-product-compliance
Lightning Source LLC
LaVergne TN
LVHW052001060526
838201LV00059B/3764